Who Am I?

Cute Pets

Written by Read With You
Center for Language Research and Development

The characters and events portrayed in this book are fictitious. Any similarity to real persons, living or dead, is coincidental and not intended by the author.

Published by Read With You Publishing

Designed by Read With You Center for Language Research and Development

ISBN-13: 978-1-944710-81-1
ISBN-10: 1-944710-81-7

Third Edition July 2019
Printed in the United States of America.

I sweat through my paws.

Who am I?

cat

I have a good sense of smell.

Who am I?

dog

I sleep with my eyes open.

Who am I?

rabbit

I can store food in my cheek and eat it later.

Who am I?

hamster

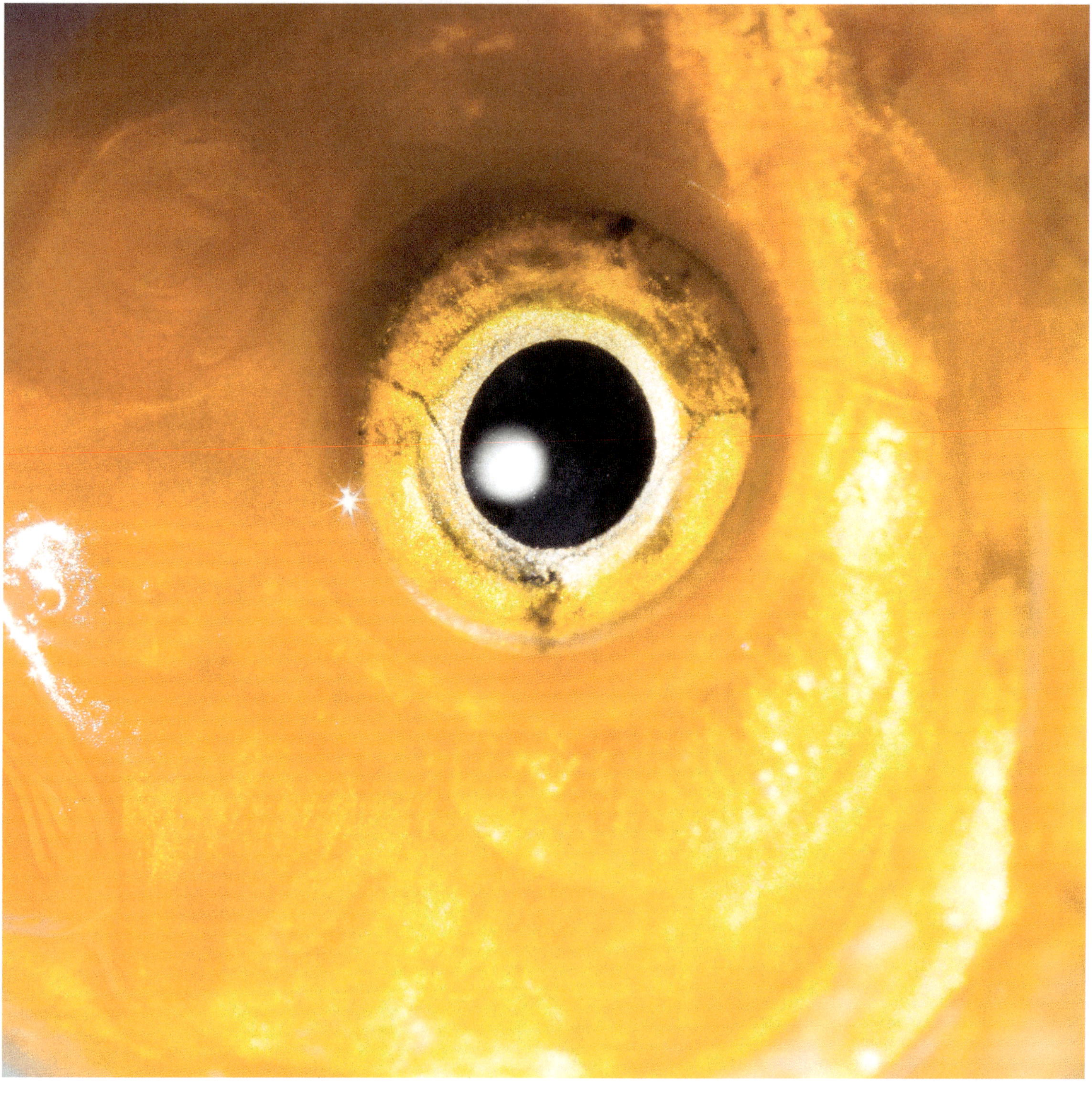

I have teeth at the back of my throat.

Who am I?

goldfish

I love to feed my partner.

Who am I?

lovebird

I blow bubbles to make a bubble nest.

Who am I?

betta fish

My teeth never stop growing.

Who am I?

guinea pig

Cat
Dog
Rabbit
Hamster

I am a cat.

I am a dog.

I am a rabbit.

I am a hamster.

Goldfish

Lovebird

Betta Fish

Guinea Pig

I am a goldfish.

I am a lovebird.

I am a betta fish.

I am a guinea pig.

We are pets.

A pet is an animal that lives with people. Pets play an important role in many people's lives.

Made in United States
Orlando, FL
07 November 2026

90337532R00017